Sharks

Blacktip Reef Sharks

by Julie Murray

Dash!

LEVELED READERS

1

An Imprint of Abdo Zoom • abdobooks.com

Dash!
LEVELED READERS

Level 1 – Beginning
Short and simple sentences with familiar words or patterns for children who are beginning to understand how letters and sounds go together.

Level 2 – Emerging
Longer words and sentences with more complex language patterns for readers who are practicing common words and letter sounds.

Level 3 – Transitional
More developed language and vocabulary for readers who are becoming more independent.

THIS BOOK CONTAINS RECYCLED MATERIALS

abdobooks.com

Published by Abdo Zoom, a division of ABDO, PO Box 398166, Minneapolis, Minnesota 55439.

Printed in the United States of America, North Mankato, Minnesota.
102019
012020

Photo Credits: iStock, Shutterstock, © Justin Gilligan / SeaPics.com p4
Production Contributors: Kenny Abdo, Jennie Forsberg, Grace Hansen, John Hansen
Design Contributors: Dorothy Toth, Neil Klinepier, Victoria Bates

Library of Congress Control Number: 2019941280

Publisher's Cataloging in Publication Data

Names: Murray, Julie, author.
Title: Blacktip reef sharks / by Julie Murray
Description: Minneapolis, Minnesota : Abdo Zoom, 2020 | Series: Sharks | Includes online resources and index.
Identifiers: ISBN 9781532129209 (lib. bdg.) | ISBN 9781098220181 (ebook) | ISBN 9781098220679 (Read-to-Me ebook)
Subjects: LCSH: Blacktip shark--Juvenile literature. | Sharks--Juvenile literature. | Fish--Juvenile literature. | Ocean animals--Juvenile literature. | Top predators--Juvenile literature. | Carnivores--Juvenile literature.
Classification: DDC 597.33--dc23

Table of Contents

Blacktip Reef Sharks

Blacktip reef sharks live in warm waters around the world.

They are mostly found in shallow waters near **coral reefs**.

They are small sharks. They measure about 6 feet (1.8 m) long. They can weigh up to 53 pounds (24 kg).

Their top halves are gray in color. Their bellies are white.

Their fins have black tips. This is how they got their name!

Blacktip reef sharks have large eyes. Their noses are round and short. They have long **gill** slits.

Blacktip reef sharks can move fast. They jump out of the water and spin.

They mainly eat fish. They will also eat squid and rays.

They are usually **timid** sharks.
But sometimes they do attack!

More Facts

- Blacktip reef sharks have been seen in freshwater areas like **mangroves** and **estuaries**.
- These sharks have rows of long, sharp teeth. Their teeth are good for catching and tearing apart prey.
- Adults often live and stay in a small area for many years.

Glossary

coral reef – a ridge of rock in the sea formed by the growth and deposit of coral.

estuary – the wide part of a river's lower end, where it meets the sea.

gill – an organ used for breathing by fish and other animals that live in the water.

mangrove – certain types of tropical trees that grow in masses along shores, with roots that grow above ground and tangle together.

timid – shy and calm.

Index

Online Resources

Booklinks
NONFICTION NETWORK
FREE! ONLINE NONFICTION RESOURCES

To learn more about blacktip reef sharks, please visit **abdobooklinks.com** or scan this QR code. These links are routinely monitored and updated to provide the most current information available.